AF681868

Planet Problem Solvers

BREAK IT DOWN

A Look at Composting

by Heather DiLorenzo Williams

NorwoodHouse Press

Norwood House Press
For more information about Norwood House Press please visit our website at www.norwoodhousepress.com or call 866-565-2900.
© 2023 Norwood House Press.
All rights reserved. No part of this book may be reproduced or utilized in any form or by any means without written permission from the publisher.

Credits
Editor: Mari Bolte
Designer: Sara Radka

Photo Credits
cover: ©FG Trade / Getty Images;page 1: ©Annie Otzen / Getty Images; page 3: ©Gannet77 / Getty Images; page 3: ©Christopher Hopefitch / Getty Images; page 3: ©Anze Buh / EyeEm / Getty Images; page 6: ©Elton.xh / Shutterstock; page 7: ©Peter Dazeley / Getty Images; page 9: ©AndreyPopov / Getty Images; page 10: ©Westend61 / Getty Images; page 12: ©Pramote Polyamate / Getty Images; page 13: ©Corey Hendrickson / Getty Images; page 14: ©RyanJLane / Getty Images; page 15: ©Everett Collection / Newscom; page 17: ©kozmoat98 / Getty Images; page 18: ©Merrill Images / Getty Images; page 19: ©Tim Grist Photography / Getty Images; page 20: ©Vicki Smith / Getty Images; page 22: ©LordHenriVoton / Getty Images; page 23: ©m.malinika / Shutterstock; page 27: ©nattrass / Getty Images; page 29: ©Justin Sullivan / Staff / Getty Images; page 30: ©Justin Sullivan / Staff / Getty Images; page 31: ©Nora Tam/SCMP / Newscom; page 33: ©Dickson Lee/SCMP / Newscom; page 34: ©NASA / Getty Images; page 35: ©ESA / esa.int; page 36: ©Strikeroot / Wikimedia; page 38: ©BirdShutterB / Getty Images; page 39: ©francoillustration / Getty Images; page 41: ©Renata Angerami / Getty Images; page 42: ©Peter Dazeley / Getty Images; page 43: ©Imgorthand / Getty Images; page 44: ©LeliaSpb / Getty Images; page 45: ©Jupiterimages / Getty Images;

Library of Congress Cataloging-in-Publication Data
Library of Congress Cataloging-in-Publication Data has been filed and is available at catalog.loc.gov

Hardcover ISBN: 978-1-68450-784-9
Papercover ISBN: 978-1-68404-742-0

353N—082022
Manufactured in the United States of America in North Mankato, Minnesota.

TABLE OF CONTENTS

CHAPTER 1

THE DIRTY TRUTH

It is a warm April morning near Columbus, Ohio. A line starts to form outside the community waste department. Many people have come to drop off five-gallon buckets. Each bucket has the words "Compost Clubhouse" printed on the side. They are filled with food and yard waste. Some people have also brought empty buckets. A 12-year-old girl named Sophie uses a shovel to fill them with what looks like dark brown soil.

Sophie is a member of Compost Clubhouse. The student-led Ohio organization provides homes and schools with a way to safely dispose of food scraps. Called composting, this process recycles food waste into a natural fertilizer called humus, or compost.

Sophie and her fellow volunteers teach their classmates and neighbors about composting. They also work at community events where they collect food waste and help distribute compost to local residents.

Compost can be added to soil to help plants grow. It is also a way to solve a big problem in the world.

Humus is a dark brown substance that looks like soil. It is filled with nutrients.

The average American creates about 5 pounds (2.3 kilograms) of trash every day. That adds up to around 108 billion pounds (50 billion kg) per year. About half of this waste ends up in a landfill. Landfills are large areas of land where trash is dumped. The land is lined with a special cloth. It protects the ground below from soaking up bacteria. Waste is dumped and covered with plastic and soil so it can **decompose**. The trash forms many layers. The layers become smaller as the waste decomposes. This makes room for more waste.

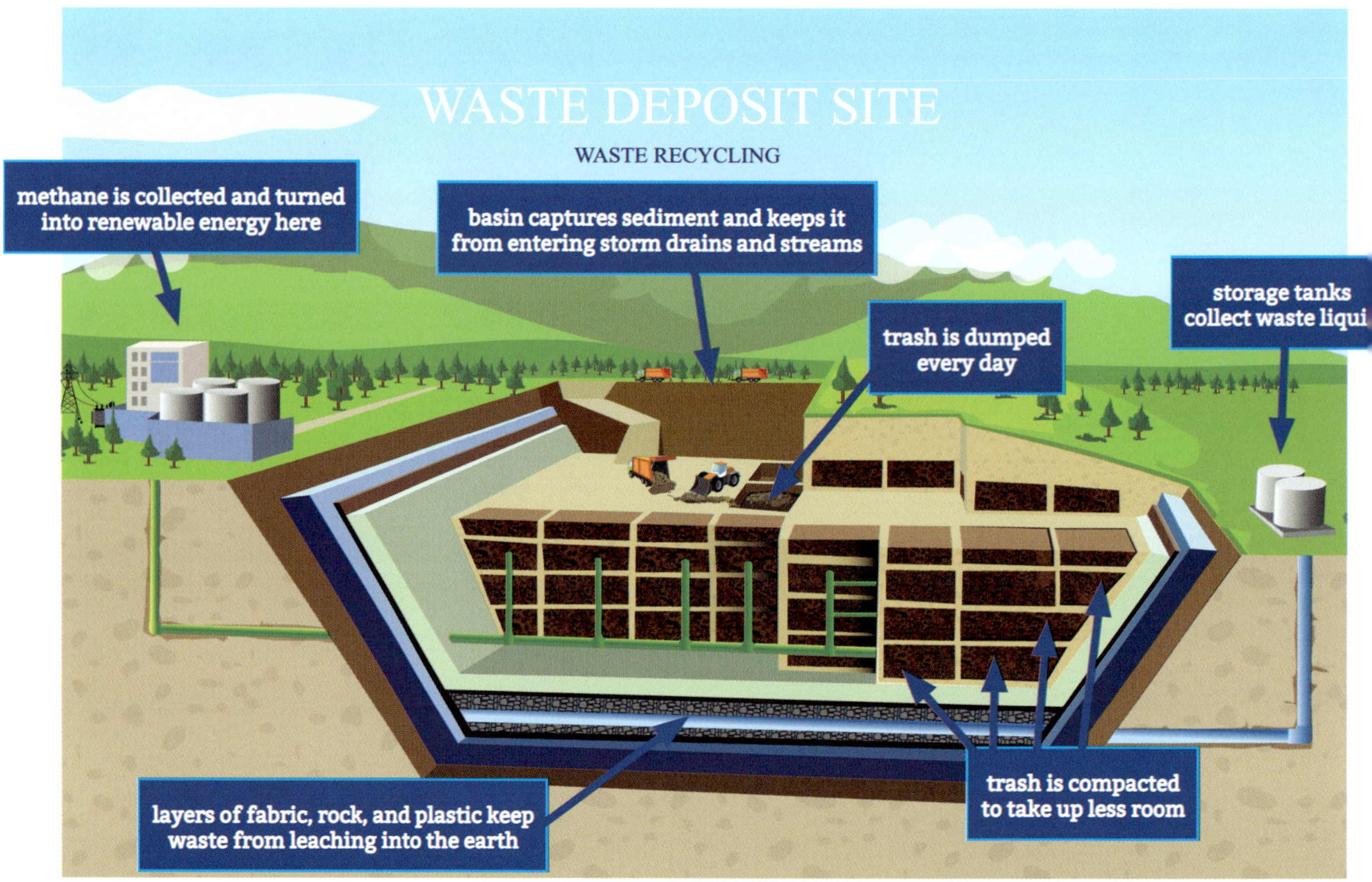

A family of four in the US throws away an estimated $1,500 worth of food each year.

More food is thrown away than any other item. Food waste breaks down quickly. But the process makes a gas called methane. Methane is a greenhouse gas. These gases absorb heat. Normal amounts of those gases help Earth stay at its ideal temperature of 59 degrees Fahrenheit (15 degrees Celsius). But the temperature rises when there is too much greenhouse gas in the atmosphere. This is known as global warming.

Methane stays in the atmosphere for around 12 years. And because so much food is thrown away, there is more methane in the atmosphere than ever. But composting can help.

Composting is a natural process. In nature, it is called decomposition. Leaves, grass clippings, and even tree branches decompose over time. Fruits, vegetables, and many other foods do too. Composting helps this process move faster. When a family chooses to compost, they are making it easier for nature to do its job.

Compost bins or heaps are easy to create. They can be in the corner of a yard or the edge of a garden. The most important part of a compost bin is what goes in it. Experts recommend a mix of "greens and browns." Green items include grass, weeds, and fruit and vegetable scraps. These make **nitrogen**. Brown items include eggshells, coffee grounds, leaves, sawdust, certain kinds of paper, and wood. These make **carbon**. Mixing these items with water and soil introduces bacteria and other **microorganisms**. Microorganisms create heat and cause decomposition.

Best If Used By

Americans throw away $161 billion worth of food each year. Around 40 percent of the food we buy ends up in the trash. One reason for this is confusing labels. Many foods have a label and date stamped on them. Some say "use by" or "best if used by." Others say "sell by." Some are not labeled at all or have only a date stamped on them. These labels are not **regulated** by the food industry. Each one has a different meaning. Unfortunately, people often throw food away based only on the date. But according to the US Department of Agriculture, some of the dates don't affect food safety. We might waste less food if the labeling system was easier to follow.

Some people throw away fruits and vegetables that are oddly shaped, even though this does not affect the way they taste.

Food waste decomposition in a landfill is an anaerobic process. That means there is not much oxygen involved. This creates more methane. Composting is an aerobic process. That means there is plenty of oxygen. This creates carbon dioxide instead of methane. Carbon dioxide is also a greenhouse gas. But the carbon dioxide made during the composting process is absorbed into the air and soil.

Do you think "best by" or expiration dates are important to food safety?

Human-made fertilizer and pesticides are applied to crops in large quantities using special farm equipment.

Plants need carbon dioxide to grow. They take the gas from the air and turn it to oxygen. This process is known as photosynthesis. It helps plants make the food they need. Farmers who grow plants for a living also use fertilizer and pesticides. They help their plants grow big and strong.

Fertilizer gives plants **nutrients** they might not get from soil. Two of these are nitrogen and **phosphorus**. Pesticides protect plants from insects, **fungi**, and weeds. Both of these products can be natural or human-made. Human-made pesticides and fertilizers cost less and are easier to use. Many come in liquid or powder form. They can be sprayed onto crops. They can be dropped from the air using planes or other machines.

Farmers in the United States have been using human-made fertilizer and pesticides since the 1940s. These products have made a big difference for farmers. Fertilizers mean bigger and better crops. Pesticides mean fewer harmful insects and diseases. Farmers grow three times more corn than they did 65 years ago. They grow twice as many soybeans.

What kinds of fruits and vegetables do farmers grow in your area? Does your family have a garden?

Safety while using pesticides is important.

Human-made pesticides and fertilizers are helpful to farmers in many ways. But they can also be harmful to humans and animals. They can enter the air, soil, and water supply. Pesticides and fertilizers are absorbed by soil. When they are sprayed, they enter the air. Rain washes the chemicals into ponds, rivers, and lakes. Eventually, they can enter the water cycle.

Pesticides are made to kill harmful insects and disease-causing fungi. But they can also hurt birds, fish, and helpful insects. Pesticides can cause skin problems if they enter the water supply. They can also lead to long-term health problems for people and animals.

Fertilizer contains nutrients that help plants grow. But over time, too much fertilizer can cause soil to stop making nutrients naturally. It can also affect water supplies. Rain washes soil and fertilizer into streams and rivers. Then, extra nitrogen and phosphorus enter the water. This causes too many microorganisms to grow in the water. They use up the water's oxygen supply. Fish and other animals can die. Other chemicals in fertilizers can make water unsafe for humans to drink.

Algae contamination can occur when too many nutrients run into lakes and ponds.

With a little hard work, food and yard scraps can turn to compost in just a few months.

But using compost instead of human-made fertilizers and pesticides can help. Composting creates a natural fertilizer. Farmers call it "black gold." Compost works with organisms such as microbes and earthworms. Together, they create a food chain that improves the soil. When it is mixed with soil, compost lets more oxygen reach a plant's roots. Compost also prevents harmful organisms that cause disease in plants.

Compost is light and airy like a sponge. It absorbs moisture and important nutrients. It feeds plants. It also acts as a filter to keep out harmful things. Compost removes chemicals from water and soil. It helps keep harmful chemicals from reaching the water supply.

Anyone can compost. Families can put food and yard scraps in a compost pile instead of throwing them in the trash. They can help keep food waste out of landfills and reduce the amount of methane in the atmosphere. Farmers can help their crops grow and protect them from pests and disease without having to use human-made pesticides and fertilizers. It only takes a little time and a few ingredients.

Rachel Carson, Earth Mother

Rachel Carson (1907–1964) was an author, biologist, and environmentalist. She grew up on a farm and loved exploring nature. She also loved to write. Carson worked for the United States Fish and Wildlife Service. There, she wrote articles for magazines and newspapers. Carson wanted people to know how much nature could be harmed by human-made pesticides. Carson wrote a book called *Silent Spring*. It talked about a world that had been damaged by the chemicals in pesticides. The book was very popular. It helped change rules about how pesticides are used in the US. Carson is considered the founder of environmental science.

What are some other items besides food and yard waste that can be reused?

CHAPTER 2

A FEW SIMPLE STEPS

People around the world have used trash dumps for thousands of years. But landfills as we know them today are **sanitary**. There are rules about how to handle waste safely. Landfills press down waste to make it smaller. They create layers to speed up decomposition. Drains and pipes help keep **toxins** from entering the water supply.

Landfills like this one are becoming too full. This is one of the many reasons composting is good for the planet.

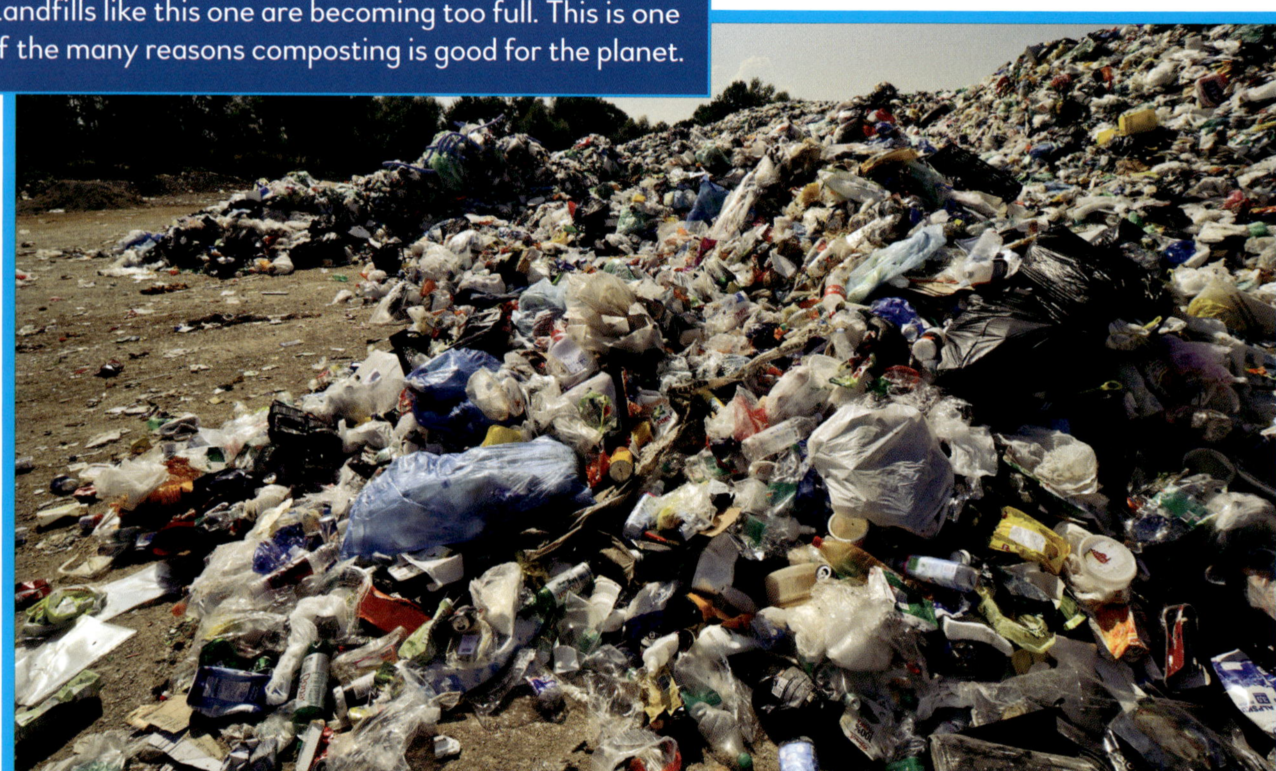

Around 50 million Americans fill 25 or more plastic garbage bags per month—that's almost a bag of trash per day!

But the US is running out of space. Some researchers think some of the country's landfills will be full within 5 to 20 years. One way to solve this problem is to send less waste to landfills.

Nearly every city and town in the country has trash collection. People place their bags of waste into large containers. Large trucks collect these bags each week and take them to a landfill. But around half of the garbage thrown away in the US can be composted.

Composting programs could reduce the waste that goes to our landfills. They are similar to recycling programs. Compostable waste goes to a compost facility. Some states have already started compost pickup.

An effective large-scale composting program is easy to follow.

The state of Vermont is one place that is leading the way. Vermont passed a universal recycling law in 2012. This means people are not allowed to throw away any recyclable material. The law applies to commonly recycled items like glass and plastic. But it also means food scraps, yard waste, and other compostable materials.

Many other US cities and states are following in Vermont's footsteps. One is California. The state plans to send 75 percent less organic waste to landfills. San Francisco has been composting citywide since 1996. It was the first city in the US to add composting to its waste collection program. It also has the highest composting levels in the country.

Other top composting cities in the US include Seattle, Washington; Boulder, Colorado; and Portland, Oregon. Portland's program includes weekly pickup of yard and food waste. The city's website gives clear instructions and helpful tips. This is one reason Portland's program is successful.

The city of Seattle provides carts for composting. Food waste, yard waste, and other organic material is picked up weekly. Leaves, vegetables, plant and tree clippings, and grass clippings are some of the things that can be composted.

Families need to be willing to separate their food scraps from their trash in order for a citywide composting program to work.

Portland's city government website is easy to understand. Lists, photos, and helpful videos provide simple instructions. Large-scale composting programs can fail if people are not sure what to do or if the process is too hard.

In 2013, the New York borough of Staten Island introduced a food waste collection program. It was part of the city's regular trash pickup. The program was a voluntary experiment. People did not have to participate. Many found separating food waste from garbage too inconvenient. The city also had trouble dealing with the food waste that was collected. There were not enough facilities in the area to keep up with all the food waste in the city. By 2020 the program was canceled.

Large-scale composting programs can be very helpful. But every household needs to participate for them to work. Also, cities must have enough money to support a separate program. And they must have a place to process the food waste once it is collected.

How many people do you know who compost? Ask around at your school, in your neighborhood, or at your next family gathering.

Sometimes, even something helpful like composting can cause problems. Large composting facilities must be carefully monitored, just like landfills. One problem comes from the waste people compost. If a homeowner uses human-made pesticides on plants or grass, they could **contaminate** large amounts of compost. And certain items cannot be safely composted. For example, pet waste or fireplace ash can ruin compost. Diseased plants can cause problems too. Some composting facilities don't get hot enough to kill weeds and diseases.

Composting facilities can also affect local water supplies. A facility in Massachusetts had more food waste than it could handle. As a result, some nutrient-rich materials overflowed. They ran into the local water supply. The water was contaminated.

Large amounts of pesticides cannot be broken down in the composting process.

Every state has rules and regulations about composting. They don't all have a large-scale program in place. But composting does not have to be done on a large scale to make a big difference.

With a few simple steps, anyone can start a compost system. Some people use a compost bin or tumbler. These can be purchased at a hardware or home improvement store. A compost bin looks like a large trash can. The compost inside must be mixed by hand using a shovel. A tumbler is on a base that spins. This helps to mix the compost up more easily. A bin or tumbler can cost $100 or more. But families can build a simple box or bin using wood or old fencing.

A bin or tumbler isn't even necessary. A corner of a lawn or garden can serve as a compost heap. Grass trimmings and leaves will compost naturally if left alone. Over weeks or months, worms and other creatures take care of the decomposition process.

Misfits Make a Difference

There are other ways to fight food waste. Misfits Market was started in 2018. A young entrepreneur wanted to make a difference. Abhi Ramesh decided to "rescue" produce. Every day, fruits and vegetables are rejected by grocery stores because they are too big, too small, or oddly shaped. Ramesh and three friends started a subscription box program. They sent the extra food to people all over the country. Over time, Misfits Market expanded. They added items that are past their "sell by" date but are still high quality. They also gather **surplus** items that would otherwise likely be thrown out. Misfits Market is dedicated to ending food waste. They also give access to quality produce and products to people around the country.

Store-bought compost bins close easily to regulate temperature and keep pests out.

Adding layers of food waste and more yard trimmings helps make compost faster. This also requires a bit more maintenance. Once food scraps are added to a compost pile, it must be turned or mixed often. Turning helps keep the scraps from smelling bad. It must also be covered and kept moist. Mixing or turning the scraps makes sure moisture is spread evenly. Worms will still help out, but the mixture of food and yard scraps creates microbes that work together to make compost.

There are a few important things to keep in mind when composting at home. Snakes, mice, and other critters are always looking for a free meal. The best way to keep them away is to make sure the compost does not smell bad. This means not adding meat, fat, or dairy to the pile. It is also important to keep the pile moist, but not too wet. Too much moisture can smell like garbage. And garbage smells like food to wild animals.

It is also important to keep chemicals out of the compost. Human-made fertilizers and pesticides from yard waste can contaminate the compost bin. Chemicals from certain kinds of paper are also bad for compost.

Finally, compost piles and bins should have the option to be covered. Even a backyard heap should have a plastic tarp over it. And keeping the compost bin in a dry, shady area helps with the temperature and moisture levels.

Even though it is simple, keeping a home compost bin or pile does require a commitment. Families may invest money in a premade bin. A bit of time and attention is necessary to make sure the compost is healthy. But these small actions are worth the big impact composting can have on the future.

Create a flyer advertising a community composting program. How would you capture people's interest? What information would you provide?

Regularly turning a compost pile reduces odors and helps regulate the temperature.

CHAPTER 3

THE KEY TO SUCCESS

The methane created by 25 million tons (22.7 million metric tons) of food waste is equal to the pollution created by 7.8 million cars. Composting that food waste instead of throwing it away could make a big difference.

Large composting programs like the one in San Francisco are proof. San Francisco composts 255,500 tons (231,786 metric tons) of food and yard waste every year. Hotels, grocery stores, and restaurants are required to compost or find other ways to reduce food waste. The city partners with farms and businesses to make the process easier. San Francisco's goal is zero waste.

The keys to San Francisco's success are city and state laws. Rules about what people can throw away are important. Giving people information and education on what to do with their waste helps programs like San Francisco's succeed. Tools and clear instructions have made San Francisco the top composting city in the country. But there's even more to it than that.

Sharing fresh produce, starter plants, and compost are all ways to work together to reduce food waste.

San Francisco composts more than 500 tons (453.6 metric tons) of waste every day. The city's compost facility is huge. Big machines sort the waste and make sure no plastics or other trash gets in. The waste is piled on the ground where it gets plenty of oxygen. Underground pipes keep the temperature stable. Employees make sure the waste gets plenty of moisture. In a few weeks, the waste turns to compost. Then, it is taken to local farms.

Compost facilities in the San Francisco area cover many acres and provide compost for local flower and fruit farms.

Composting machines speed the process up and help large businesses process food waste much faster.

But a company in China invented a machine that can do all of that in 24 hours. Now they are available in the US. These devices take up less space than huge compost yards. Some are made for schools, hospitals, hotels, airports, and other places where food is made. They are large enough to process a whole day of food waste. The machines have an opening for discarded food. Special microorganisms are added to the machine. A set of paddles rotates the contents constantly.

The machines are temperature-controlled. This kills any harmful bacteria. The food waste turns to compost in a day. Depending on their size, they can treat between 10 and 500 tons (9 to 454 metric tons) of food waste per year.

These same machines are also available for smaller businesses. Even families can buy them. Some are about the size of a trash can. Others are small enough to fit on a countertop. These machines are good for families who do not have yards or enough space for a compost heap.

The smaller machines use the same process as the larger ones. They make sure the temperature is just right. A microbe tablet is added to the machine once a year. Like the large machines, the waste is processed in 24 hours. These machines can process up to a ton of food waste each year.

Although these machines cost a lot, they are an example of how technology could change the way people compost. Scientists and inventors are looking for ways to reduce greenhouse gases. They already know that keeping food waste out of landfills is part of the solution. New inventions like these machines could make composting easier.

Schools throw away around 36.5 pounds (16.5 kg) of food per student every year. What ideas do you have to help your school waste less food?

No matter whether it's low-tech or high-tech, composting can make a big difference.

Astronauts like Scott Kelly (pictured here) must be creative about how they dispose of waste on the ISS.

One place where new inventions are always being tested is outer space. Many things we use every day got their start on the International Space Station (ISS). Memory foam, medical devices, and even water filter systems were created by or for astronauts.

Astronauts have been recycling on the ISS for years. Astronauts even have a way to turn their urine into drinking water! The National Aeronautics and Space Administration (NASA) is currently experimenting with plastic recycling. They turn their plastic waste into materials that can be used for 3D printing. Then, they can make new things that they need.

NASA is also exploring ways to avoid food waste in space. One uses **fermentation**. This is a very specific kind of composting. It can compost all kinds of food waste, including meat and dairy products. Using fermentation on the ISS could give astronauts a way to preserve some foods. They would store them in bags along with microbes. The bags could also be used to grow certain foods and process food waste. NASA is thinking ahead to possible life on Mars.

They Recycle *Everything*

Around 30 years ago, scientists from the European Space Agency (ESA) began researching ways to keep astronauts alive and healthy in space. They wanted to create an artificial ecosystem that mimicked life on Earth. Micro-Ecological Life Support System Alternative (MELiSSA) is part of their research. The goal of MELiSSA is to create a circular life support system that uses all waste, including bodily waste, to create water, food, and oxygen. MELiSSA is made up of compartments that complete the same steps that occur in the natural life cycle. All waste from a space mission is turned to liquid. Then, it is turned into useful gases. The next step is similar to photosynthesis. Finally, the waste is turned into a food product. MELiSSA has been tested with rats, but humans are next up in the research process!

Many cities have food and yard waste drop-off stations in community gardens.

Not every city in America requires people to compost. And not everyone can afford a special composting machine. But every state has resources for residents about how to waste less food. Many have guides about how to compost and where people can learn more.

In 2020, the United States Department of Agriculture (USDA) began offering **grants** for community composting programs. These grants total around $2 million each year. In 2021, 24 cities and towns received money to start compost and food waste reduction programs.

These projects include community compost stalls and food scrap pickup programs. Stamford, Connecticut, purchased a large composting machine. It is kept at a local science museum. Residents can bring their food waste to the museum to be composted. In Philadelphia, Pennsylvania, the Parks and Recreation Department set up food waste composting stations in city parks. And in Richmond, Virginia, food waste sites were placed at libraries, small businesses, and community centers.

Government grants help cities and communities start programs and find solutions. Kids are doing their part too. Youth-centered programs in Baltimore, Maryland, and Columbus, Ohio, are proof that kids can change the future.

Incinerators release harmful chemicals and pollutants into the air.

Baltimore was once one of America's 20 most polluted cities. Baltimore uses a large trash burning device called an incinerator to dispose of much of its waste. In 2005, more people in Baltimore died from illnesses caused by air pollution than any other city in the US. The city knew it was time to turn things around. People in Baltimore have been working hard to slow down or stop pollution. This includes burning less trash.

A group called the Baltimore Compost Collective is helping. Marvin Hayes founded this youth-led organization. Hayes wanted to help reduce the amount of waste that was taken to Baltimore's incinerator. He also wanted to prepare young people for the future.

Hayes and his teen volunteers collect food scraps from residents in five major Baltimore neighborhoods. They take the waste to a compost pile at the Filbert Street Community Garden, which Hayes also manages. The teens learn how to sort the food waste. They learn about composting and gardening. They also learn how to work with others and with the public.

The Baltimore Compost Collective keeps around 500 pounds (227 kg) of food waste from the incinerator each week. These teens are helping Baltimore become a cleaner, greener city.

Food waste starts in the kitchen at your house. It is collected, sorted, and turned into compost. The methane made during this process is used to make renewable energy. Then, the compost is used as fertilizer for the food you buy and eat. Then the process begins again!

A few states over in Columbus, Compost Clubhouse is also making a difference. The group is made up of kids ages 10 and older. With the help of adult volunteers, they lead composting programs in their schools and neighborhoods.

Compost Clubhouse was founded in 2019 by Mona Barber. She knew it was important to reduce food waste in landfills. But she also knew it would take a long time for her community to pay for a composting program. So Barber and her kids started their own.

Today, around 170 kids are part of Compost Clubhouse. The volunteers receive a business card and a staff T-shirt when they join the group. The experience not only helps the Columbus area reduce food waste. It also gives the young volunteers important life skills and work experience.

Even though shoveling compost is hard work, Sophie knows she is making a difference. She joined the group after visiting a landfill on a class field trip. Sophie and the other volunteers learn about how composting helps the planet. They share what they know with classmates and neighbors. Through these small actions, they are making big changes. In the words of Riley, another Compost Clubhouse volunteer, "By composting, we are creating a world for our friends, family and future children to live in . . . one corn cob or strawberry stem at a time!"

Groups like Compost Clubhouse and others across the country teach friends, families, and classmates about the importance of keeping food out of landfills.

Research volunteer opportunities in your town. Are there any that sound interesting to you? Share them with a friend. Then, find out how you can volunteer yourself!

Composting might take some getting used to, but it's a habit that can make a huge difference.

People in the US throw away approximately 40 percent of the country's entire food supply every year. From table scraps to expired products, the average person wastes about 200 pounds (90.7 kg) of food yearly. Most of it gets dumped in a landfill. It rots there and creates methane. But there are ways to keep this from happening. Composting is one option.

Composting is a natural, easy way to make something new from our food waste. It can be as simple as a pile of leaves, coffee grounds, banana peels, and eggshells in the backyard. It can be as complicated as a high-tech machine that turns waste to compost overnight.

Some people live in communities where composting is required. Others have teams of future-minded kids who are determined to change what we do with our trash. One school, one family, one neighborhood, and even one kid on a mission can make a huge difference through composting.

Countertop compost bins can be used to collect kitchen scraps. When they are full, the scraps can be added to outdoor piles or bigger bins for pickup.

Activity 1:

Start collecting your own food scraps. Create a small compost heap in your yard or garden. If you don't have a yard, find out if there's a community garden nearby. You can take your scraps there.

Activity 2:

Take your compost bin to the next level. Start a compost program at your school. Talk to your teacher or the principal and ask if a special compost bin can be placed in the cafeteria (this can even be a trash can with a sign on it that says "compost only").

Make signs with your classmates to show your fellow students what goes in the compost bin. Refer to page 23 to get started. Research other items that can be composted.

Then, create a compost pile at your school!

GLOSSARY

CARBON (KAR-buhn): an element that easily combines with other elements to make chemical compounds

CONTAMINATE (con-TAM-uh-nayt): to make dirty or poisonous

DECOMPOSE (de-kom-POSE): to rot

FERMENTATION (fer-men-TAY-shuhn): the process of breaking down into a simpler substance

FUNGI (FUNG-eye): organisms that often live off of other organisms, such as mold, mildew, and mushrooms

GRANTS (GRANTS): a sum of money given to someone in order to fund a project or solve a problem

MICROORGANISMS (mi-kroh-OR-guh-niz-uhms): living things that are too small to be seen with the naked eye

NITROGEN (NY-tro-jen): a colorless, odorless gas that makes up most of Earth's atmosphere

NUTRIENTS (NOO-tree-yunts): chemical compounds used by living things to function properly

PHOSPHORUS (FOS-for-us): an important element that is essential for living things

REGULATED (REG-yoo-layte-uhd): to set rules about how a business or procedure should operate

SANITARY (SAN-uh-tare-ee): clean and healthy

SURPLUS (SUR-plus): extra

TOXINS (TOKS-ins): poisonous substances that can hurt living things